COLORADO

CRAIG AURNESS & RICHARD D. LAMM

BOULTON

COLORADO

Designed by Fortunato Aglialoro
ISBN 0-920831-01-X
1 2 3 4 8 9 7 6
Printed in Hong Kong
by Scanner Art Services Inc., Toronto

INTRODUCTION

Colorado is a state with many blessings. It has unsurpassed scenery, great stores of natural resources, and a fascinating western heritage. Its geography is spectacular. Rolling plains give way to gentle foothills, which in turn give way to rugged mountain ranges filled with valleys, basins, passes, and river bottoms. The Continental Divide runs north and south through Colorado and the topography, like a kaleidoscope, changes with every mile.

The word 'Colorado' came from the Spanish adjective meaning 'red' or 'ruddy', but that word clearly does not do justice to the yellows of the aspen, the green of the conifers, or the blue of the horizon. Simply put, Colorado is dramatic, its beauty is majestic and its vistas are endless.

The first people to live in Colorado are lost in history. We know that by the start of the Christian era basketmakers were living in the Mesa Verde area of southwestern Colorado. Today, driving from Durango to Cortez, one can see the table-land country that was home to these Anasazi Indians. As the basketmakers' culture developed into that of the modified basketmakers and then into the pueblo period, we find a dramatic increase in the sophistication of their domestic lives and their construction techniques. By AD 1100, the native people of Colorado were building apartment houses that were so large they were not equalled by modern methods until the 1880s. Their ruins haunt us to this day with many unresolved questions: who were these people, where did they go, who did they trade with, what was their lifestyle? They had no horses and they had no livestock. They used no metals, but they were so ingenious and imaginative that they achieved a high quality of masonry work. They dominated and held this area for hundreds of years. To this day we are in awe of their accomplishments. Their Cliff Palace, for instance, contained 200 one-room houses, a city within a city for approximately 400 people.

For unexplained reasons, toward the end of the 13th century, they left their cliff houses and migrated elsewhere. Historians speculate that their leaving had a great deal to do with the drought years that lasted from 1272 to 1299. Dependent on agriculture, they were unable to maintain their livelihood in the face of extended drought. We do know that by the beginning of the 14th century, hundreds of square miles of the area contained ancient ghost towns, the inhabitants gone, their structures partly protected against the ravage of the elements by the cliffs in which they had sought shelter and protection.

Gradually, other Indians moved into Colorado from a wide variety of different areas, and when the Spanish arrived in the 17th century they found a number of tribes already inhabiting Colorado. One of the best known of the Spanish expeditions was led by two Franciscan priests, Dominquez and Escalante who, in the historic year of 1776, travelled a wide area of western Colorado. These two intrepid priests explored the San Juan Mountains, the Dolores River, the Gunnison River, and then turned west on into Utah. But settlers did not follow them. For reasons that we are not quite sure of, although the Spanish colonized much of New Mexico, they did not settle north of Santa Fe, New Mexico. Thus Colorado remained isolated and almost undiscovered. The native peoples still had Colorado freely to themselves.

A half a world away events were taking place that would forever change their lives and their culture. In 1800, Napoleon Bonaparte acquired Louisiana from Spain, in hopes of reviving his French empire. When these dreams failed, he sold Louisiana to the

United States and Thomas Jefferson quickly set out to find what it was that he had bought. While Lewis and Clark were still on their journey of discovery, another expedition was dispatched toward the southwest under the command of Zebulon Montgomery Pike. Leaving a fort near St Louis on 15 July, 1806, Lieutenant Pike headed west, almost totally ignorant about what he would find there. On the 15th day of November near the present town of Las Animas, Pike saw what looked like a low cloud, but which he soon found to be what we today call 'Pikes Peak'. What looked like a one-day journey to the base of the peak turned out to be a two-week march. Upon reaching its base, Pike became the first of hundreds of thousands since who have been unable to resist the lure of trying to climb to the top. He failed. But the Rockies had been discovered, and the pace of change in this area has gone on increasing geometrically ever since.

Pike was soon followed into the West by some of the more fascinating characters that our history has produced: the mountain men. A last vestige of total freedom, these rough characters came west looking for the beaver and left a heritage that is with us to this day. Fiercely independent, ruggedly self-reliant, a thousand miles from the nearest hardware store, they had to live completely off the land. Their doings are part of our heroic legends like the story of Pegleg Pete, who had to amputate his own leg! The Bent Brothers, who built the first fort in Colorado along the Santa Fe Trail, arrived to make their mark both on Colorado and New Mexico history. These mountain men found the passes, marked the trails and waterholes. They were the harbingers of the civilization that was soon to follow. Their forts (St Vrain, Bents Fort, Davey Crockett) were Colorado's first permanent settlements. Truth be told, it was more the nomadic life and their own iconoclasm that brought them west. The beaver was often secondary. But be that as it may, they came out west and they seldom went back. American history is full of such fascinating, self-reliant, nomadic characters who could not stand to see any kind of 'civilization' between them and the sunset.

Civilization came with a rush to Colorado, starting in the summer of 1858, when rumors abounded throughout the East and Midwest of gold discovered near Denver City. Whether these early rumors were true or false, we do know that on 6 May, 1859, William Green Russell of Georgia found gold and the rush was on. Gold was the magnet that peopled early Colorado.

Whenever people are attracted to a place they cluster there and in Colorado, in 1858, the cluster was at the confluence of Cherry Creek and the South Platte. Originally three different towns, St Charles on the east bank of the river, Denver City north of St Charles, and Auraria on the west side of Cherry Creek, this location was soon given the name 'Denver' (after the Governor of Kansas Territory, of which it was then a part) and Denver soon became a bustling city. An early trickle of people soon turned into a flood as this 'new California' continued both to rise and then to shatter dreams. Any place there was 'color' a mining camp would spring up and many of Colorado's present towns, like Idaho Springs, Central City, and Blackhawk, had their beginnings in those days.

Though Colorado's gold rush produced far less gold than California's gold rush, it did establish a permanent civilization, both along the Front Range and on up into the mountain valleys. For the next 30 years, until the silver crash of 1893, mining dominated Colorado's history. Fired with hope, legions of men and women tramped across the Kansas prairies to find a chance to strike it rich. Their stories are part of the warp and woof of our western history. The line between success and failure was often cruelly narrow. A miner would look for years in vain for a strike. Then his mule might accidentally kick an outcropping, only to uncover a phenomenal strike of silver. After looking fruitlessly for gold, a man might decide to build a house and, while digging his basement, discover a mother lode. With a capriciousness that defies analysis, a few got rich, but for every success story, there were a thousand broken dreams. 'Pike's Peak or Bust' was crossed off the side of the wagon and 'Busted by God' written there instead by thousands of people who came to be known as the 'go-backers'.

But many did stay, converted not by gold, but by the magnificence of the landscape and the fruitfulness of the land. The early cattlemen brought large herds to feed the miners and later to feed America. Denver City became just plain Denver and hundreds of mining towns dotted Colorado's landscape. Soon the day of the individual prospector turned into the day of the large corporation which hired men to work deep underground in almost unbearable conditions. The independent prospector years in Colorado passed in a wink. Mining became a big corporate enterprise, fraught with labor strife and industrial takeovers.

Towns like Buckskin Joe, Tincup, Town Tomboy, Royal Flush, Telluride, Silverton, Silvercliffe, Bonanza, Eureka, and Tarryall, exploded onto the landscape all filled with bustling people who dreamed big dreams. These people risked starvation, avalanche, freezing snow, Indians, and outlaws in seeking wealth from the earth. As the early gold strikes petered out, they were soon replaced by the discovery of silver. Phenomenal fortunes were made all over the state by silver miners. Then in 1893, at about the time that the silver busted, Cripple Creek was discovered by Bob Womack, a cowboy chasing a crippled cow. Cripple Creek soon became one of the world's largest gold-producing mines, adding phenomenal wealth to the national economy during its years of operation. To this day, mining is not only an important part of our heritage and history, but an important part of our economy. At the same time that the miner infiltrated the mountains of Colorado, another bigger-than-life historical character, the cowboy, established himself on the eastern plains. The classical movie cowboy was a myth. In fact a cowboy's job was one of the hardest, dirtiest, most dangerous jobs in all America. Willing to ride hundreds of miles a day, day in and day out, risking storms, sleet, rain, hail, dust, temperatures over 100 degrees, sleeping on the ground with the saddle as a pillow, the cowboys toiled long hours for very low pay. But their legend lives on in America, perhaps stronger than any other. The cowboy became a symbol of the self-reliance, the righteousness and the ambition of America. To this day, our movies, TV drama, footwear, headgear, and part of our language are dominated by the memory of the cowboy.

At the same time that the cowboy was becoming an established character, a less glamorous but more numerous type of occupational cousin was putting down roots throughout Colorado. The homesteader or the sodbuster brought permanent civilization. In the San Luis Valley, both Hispanic and Anglo farmers gobbled up the empty land and planted their roots. Into the mountain valleys, Colorado's Front Range, and out on the eastern plains, the sodbusters came in large numbers. Dauntless souls who would risk hail, drought, grasshoppers, freezing snow, and searing sun, they built farms and ranches throughout Colorado. It is impossible to overestimate the lure that 'free land' had on the immigrant. Railroads would hype their lands in newspapers throughout Europe, and hundreds of thousands of landless hopefuls streamed into the West and into Colorado, seeking the promise of free land. When they got there, the soil was often not as rich as had been promised and there was often no timber even to build a house. They had to live in dugouts or sod huts. Winter would come with a fierceness that drove temperatures far, far below zero. The wind whipped the snow so that it actually snowed inside the sod huts. Cattle succumbed by the thousands to the winter weather. Spring brought grasshoppers. Summer brought drought and then hail. Flash floods would turn dry gullies into raging torrents. On and on they came, telling themselves that the 'rain would follow the plow'. Life was unbelievably hard, but the settlers were willing to put up with all of this to have their own land in an empty continent. But many a bright dream turned to cruel nightmare in this unforgiving land.

Some arrived in communal groups, like the Union Colony in what is now Greeley, Colorado. This new colony, organized in 1869, accepted any one of 'good character who was a temperance advocate and who had $5 to be initiated and $150 for dues'. In return, a member was entitled to a piece of farmland from five to 80 acres in size (depending on the distance from town) and had the right to buy a lot in the new town that became Greeley. The Colony bought approximately 12,000 acres of land and then individual members filed on some 60,000 acres of public land that surrounded the settlement site—'to prevent uncongenial neighbors'. Hundreds of people, following Horace Greeley's historic advice to 'go West' joined the organization and arrived in Union Colony to settle their roots into this new settlement. Within the first two months, Greeley boasted 130 buildings and by the spring of 1871 it had filed incorporation papers, elected officers, and started to develop the surrounding farmland. Greeley grew and prospered.

As agriculturalists came into this semi-arid climate, they had to develop irrigation. Water was diverted by canals from the South Platte and gravity distributed it through irrigation ditches to the fertile, porous land. Ingenious irrigation often characterized the development of Colorado agriculture, the systems usually being built by cooperative groups. One poet observed that, 'The history of the West is written in water'.

While all of these groups, mountain men, miners, homesteaders, sodbusters, cowboys, were lured to Colorado for economic opportunities, suddenly there came a new type of visitor who was not interested in making a living but in enjoying the scenery; the age of the tourist had begun. In 1867 Bayard Taylor published an extremely successful book called *Colorado, A Summer Trip,* which told of the scenic and recreational joys of Colorado. Resorts started in places like Colorado Springs, Glenwood Springs and Estes Park. Tourism was often a by-product of the railroads and soon even the Narrow Gauge D. & R.G. was taking tourists up into the old mining camps to view the rugged scenery and savor the clear mountain air. Tourists grabbed the sides of the railroad cars with white knuckles as the trains rattled over wooden trestles, thousands of feet above raging mountain torrents, and groaned up the high-country passes. The Georgetown Loop was an unforgettable train ride, accessible to Denver, which soon became a great popular favorite. A series of Victorian hotels sprang up to serve the tourist. Health spas like the Hotel Colorado at Glenwood Springs were built on a scale seen before only in Europe.

While all this was happening, Denver was undergoing profound changes. From being a frontier settlement of log cabins, it soon blossomed into a Victorian city. As the wealth poured out of the mountains and out of the plains, the *nouveaux riches* moved to Denver and soon made it a place of pretentious homes, glittering gambling houses and plush brothels. No architectural style was too fanciful for Denver. Gothic castles, with Italian turrets, Victorian houses, gingerbread storefronts—the unsinkable Mrs. J.J. Brown built a house of lions, with two sphynx-headed lions to guard it.

Into this 'Queen City of the Plains' strode an incredible assortment of people—miners, merchants, ex-cowboys, and Civil War veterans. Here came 'Baby Doe', a *divorcée* who married a bonanza king and died keeping vigil over his 'Matchless Mine' long after his death; here came the saints and here came the sinners, the entrepreneurs, and the immigrants, looking for their place in the sun. World War II brought a new wave of people to Colorado and transformed Denver into a cosmopolitan city.

Colorado is only ten years into its second century of statehood but this 'Centennial State' has come a long way.

The state and its 'Queen City of the Plains' are both alive with a bustling vigor and synergism. Colorado has a thousand faces: the businessman in Denver creating an empire, the former steelworker in Pueblo now training for a new high-tech job, the cattle rancher on the eastern plains, the potato farmer in the San Luis Valley, the miner in Creede, the motel operator in Durango, the ski-lift operator in Vail, the retired millionaire in Aspen, the Hispanic artist, and the black judge. Colorado is alive with many cultures and many viewpoints. A young, dynamic, Hispanic mayor presides over its major city. A rural rancher presides as Speaker of the House. A woman sits on the Supreme Court. And for all their diversity, the people of Colorado are united in the unshakable confidence with which they look to the future.

RICHARD D. LAMM, GOVERNOR

1 Skiing in the Aspen Highlands.

2 *(left)* In Steamboat Springs, visitors are able to take rides in a hot air balloon. In the background is the Steamboat Springs Ski Mountain. Steamboat Springs is one of the major ski resorts in Colorado.

3 The Telluride Hang Glide Festival, held in autumn, offers a spectacular setting for this dangerous but awe-inspiring sport. Favorable wind currents can offer a skilled glider rides lasting for hours; a sudden change in atmospheric conditions can make this location too dangerous for experts to attempt. Sponsored by a local group known as the 'Telluride Air Force', this competition in the former mining town of Telluride is one of the most popular with hang gliding enthusiasts. The trail of smoke allows spectators to follow the hang glider's maneuvers more easily.

4 In the mountains behind Aspen Ski Resort are some of the most beautiful wilderness ski areas in the country. Parts of this wilderness ski area can be reached only by off-road transportation, such as snow cats. The untamed beauty and natural ski slopes attract skiers from both the United States and Europe. This skiier is on the Little Annie Ski Hill.

5 *(right)* Facing the setting sun in the mountains southwest of Aspen. This range includes Maroon Peak and Castle Peak, both over 14,000 feet high.

CLARKLIFT
CLARK
PETROLEUM

6 *(left)* The Denver skyline facing south at sunset, over Interstate 25. Denver, one of the fastest-growing cities in the United States, is considered by many to be the energy capital of the country, a title also claimed by its close rival, Houston.

7 Denver and the surrounding suburbs are models of creative new architecture. Harlequin Plaza in Englewood, Colorado, home of Travelers Insurance, is an example of the new style.

8 The aspen trees in autumn make Colorado famous for autumn color.

9 *(right)* Piney Lake in the wilderness above Vail, part of the Gore Range.

10 *(left)* Mesa Verde National Monument, one of the best examples of prehistoric Anasazi cliff-dwellings. Cliff Palace is the best-known and most dramatic of the Mesa Verde ruins. It is not completely understood why these dwellings were suddenly abandoned about 700 years ago, but radical changes in the climate are suspected as a major factor.

11 Spruce Tree House in Mesa Verde National Monument in Southwestern Colorado.

26205
27
40897
3985
72656

12 *(left)* Hobie Cat Races at Dillon Reservoir in the mountains west of Denver. Sponsored by Rocky Mountain Marine in Denver, this annual event draws enthusiasts from around Colorado and the West. 230 boats competed in these races.

13 Annual hot-air balloon festival in Colorado Springs.

14 Above Maroon Creek in the mountains southwest of Aspen.

15 *(right)* The Cadet Chapel of the United States Air Force Academy, Colorado Springs, Colorado. Hosting services in many faiths, the chapel serves the spiritual needs of future Air Force officers. Its 150-foot spires are a noted example of contemporary architecture.

16 Steamboat Springs has kept much of its Western heritage. This storefront display greeted the Steamboat Springs 69th Winter Carnival.

17 *(right)* The Greeley Independence Stampede, an annual rodeo in Greeley, Colorado, is one of many ways Colorado celebrated the Fourth of July.

18 A last stand of aspen leaves in late autumn along Maroon Road shows a genetic variant; surrounding trees have already lost their leaves.

19 *(right)* Storms are an ever-present threat to the autumn harvest. Fire, wind or rain before the crops are gathered can ruin a promising harvest.

20 *(left)* The famous clock tower at the end of 16th Street in downtown Denver is reflected off one of the newer steel and glass buildings.

21 Larimer Square, a restored historic section near downtown Denver, has many interesting shops and restaurants.

22 *(left)* The city and county building at the Civic Center in downtown Denver.

23 Construction workers on the 31st floor of a new building in downtown Denver.

24 A view north from Colorado National Monument, just west of McClure Pass on Highway 133, near Grand Junction.

25 *(right)* Looking north over Denver.

26 and 27 *(right)* C.F. and I. Steel Corporation mill outside Pueblo, Colorado. This company was one of the major manufacturers of the steel rail for the railroad, and as such was instrumental in the development of the West.

28 *(left)* Twilight in Aspen, one of the foremost ski resorts in Colorado. A former mining town, Aspen keeps much of its 'Old West' flavor.

29 Little Annie Ski Run, Aspen.

30 Aspen trees near Vail, along Gore Creek.

31 *(right)* Great Sand Dunes National Monument, one of the largest and most spectacular sand dune formations in the United States. The Sangre de Cristo Mountains are in the background.

32 *(left)* and 33 Telluride, Colorado, a former mining town, now a popular resort and retreat from big-city living for many new Colorado residents.

33 One of many restored Victorian homes in Telluride.

34 *(left)* The Sangre de Cristo Mountains near Creston Needles, above Great Sand Dunes National Monument.

35 Facing north over Great Sand Dunes National Monument with the Sangre de Cristo Mountains in the background.

36 *(left)* Monfort Feed Lots near Greeley, Colorado make up one of the largest cattle feeding operations in the country. Feeding 70,000 head of cattle, with a capacity for 100,000 head, Monfort draws upon contemporary research and technology in its operation. The weight of each individual cow and steer is tracked and scientifically-formulated feed is prescribed individually.

37 XV Congreso Charro, held at the Charro Arena near Pueblo, Colorado. In this Mexican-style rodeo, competing teams from Phoenix, El Paso, Tuscon and Pueblo participate.

38 Aspen, Colorado, with many Victorian-style homes.

39 *(right)* A cabin behind Aspen.

LITTLE ANNIE HUT

40 *(left)* and 41 Royal Gorge near Canon City has a spectacular view of the Arkansas River, and is a major tourist site.

42 *(left)* Vail, Colorado, a popular ski resort developed in a contemporary European architectural style.

43 A home in Vail around Christmastime.

44 A farm in the San Juan Mountains near Telluride, an area where there is a great deal of sheep and cattle ranching.

45 *(right)* The Durango-Silverton Railroad, one of the few remaining narrow gauge railroads in the country. It is now a popular tourist attraction, connecting the old mining towns of Durango and Silverton, which now thrive on the 'Old West' tourist trade. Originally this line served the mining industry.

481
DURANGO & SILVERTON

46 Wheat harvesting near Akron in the open prairie of Eastern Colorado.

47 *(right)* Miles of farm roads across the prairie northeast of Akron, Colorado.

48 The Maroon Bells, near Aspen.

49 *(right)* The Garden of the Gods, an attraction outside Colorado Springs. Pikes Peak is in the background.

50 *(left)* University of Colorado in Boulder.

51 University of Colorado's Homecoming Game against University of Oklahoma.

52 Rancher Jim Patrick is cutting through ice so that his cattle can get to the water to drink. Patrick and his wife and three children moved to Colorado from Oklahoma, where the family owned a bank. Now full-time Colorado ranchers, they raise cattle and horses at Steamboat Springs.

53 *(right)* A cold morning outside Steamboat Springs.

54 A rainbow above University of Colorado in Boulder, after a summer thundershower.

55 A view east over the San Luis Valley, near Alamosa.

56 *(left)* Summer golf on a course in Vail.

57 Winter near Aspen.

58 Ski-jumping into a heated pool at the Winter Carnival, Steamboat Springs.

59 *(right)* An aerial view facing north over Crested Butte, a historic mining town which has become a popular resort.

60 Rancher Calvin Powell and his son. Powell and his father-in-law, Jerry More, operate the ranch More has owned since 1948. Much of the original More ranch was sold to the developers of the Steamboat Springs resort. The remaining land they operate much as was done before the resort was developed, now renting some of the grazing land.

61 *(right)* Feeding cattle in winter near Steamboat Springs. Jerry More lives on the ranch with his son-in-law Calvin Powell and his daughter and two grandchildren.

NO 58
LIQUORS
Coors
COMPANY STORE

62 *(left)* The main street of Crested Butte, Colorado.

63 The office of the *Crested Butte Pilot,* a local newspaper. Though the town is small, it has two rival newspapers.

64 *(left)* Cross-country skiing in the Maroon Bells.

65 The mountains above Aspen.

66 Robot arms construct a computer hard disc in the clean room, Storage Technology Corporation, Boulder.

67 The Durango-Silverton railroad.
(See also plate 45)

68 *(inset)* Wheat harvest in the Eastern Colorado prairie near Akron. Changing weather causes some farmers to harvest all night, before the rain can dampen the grain and cause rot.

69 Sunset near Burlington, Colorado.

70 and 71 Fourth of July celebration at Estes Park, Colorado. The Rocky Mountain National Park is in the background.

72 Steamboat Springs Winter Carnival. Teams of torch-bearing skiers light up the hillside.